FLORA OF TROPICAL EAST AFRICA

VALERIANACEAE

J. O. Kokwaro

(East African Herbarium)*

Annual (occasionally biennial) or perennial herbs, rarely subshrubs; often with strongly scented rhizomes. Leaves opposite or sometimes forming basal rosettes, exstipulate, often pinnately much divided but sometimes entire; cauline leaves sometimes few, small or none; basal leaves pinnatifid or once-or twice-pinnatisect; base often sheathing. Inflorescence a many-flowered compound dichasial cyme, thyrse or monochasium, sometimes condensed and capitate, bracteate and usually bracteolate (the subtended bracteoles often mistaken for the calyx). Flowers hermaphrodite or unisexual by abortion (plants then dioecious, as in some *Valeriana* spp.), irregular or almost regular, usually 5-merous. Calyx often small or obsolete at the time of flowering, sometimes enlarging as the fruit matures, and then variously lobed; lobes often forming a pappus. Corolla funnel-shaped or tubular, often attenuated at the base, generally basally spurred or saccate, wider at the throat; limb 3–5-fid, oblique or 2-lipped; lobes imbricate. Stamens 1–4 epipetalous, alternating with the corolla-lobes; anthers versatile, 2- or 4-lobed, 2- or 4-thecous, introrse, dehiscing longitudinally; pollen grains tricolpate, echinate. Ovary inferior, tricarpellate, 3-locular but only 1 locule fertile; ovule solitary and pendulous, anatropous; style single and slender; stigma subtruncate, entire or shortly 2–3-lobed. Fruit a 1-seeded achene. Seed pendulous; embryo straight, large; cotyledons oblong; radicle superior; endosperm very thin or absent.

A family of about 13 genera (only 4 genera, namely *Centranthus, Fedia, Valeriana* and *Valerianella* occurring in Africa), and about 400 species occurring mostly in the temperate zone of the northern hemisphere, very common in western North America and the Andes. In East Africa, the few species known are generally confined to the high mountains.

Leaves predominantly pinnatifid or pinnatisect; calyx inrolled in flower, regular in Flora area; corolla saccate at the base; fruit unilocular; perennial herbs	1. **Valeriana**
Leaves predominantly entire; calyx obsolete, unequally toothed (oblique and entire in Flora area); corolla not saccate at the base; fruit trilocular; annual herbs	2. **Valerianella**

1. VALERIANA

L., Sp. Pl.: 31 (1753) & Gen. Pl., ed. 5: 19 (1754)

Perennial herbs, vines or subshrubs, usually glabrous or sparsely hairy with short simple hairs, with a bitter taste and peculiar smell especially when dry, fleshy (woody or tuberous) roots, sometimes stoloniferous, occasionally gynodioecious or polygamodioecious. Leaves petiolate to nearly sessile,

* Account prepared as part of a course of studies at the University of Uppsala, Sweden.

exstipulate; radical leaves entire or toothed, in basal rosette, often long-petioled; cauline leaves pinnatifid, or once–twice-pinnatisect. Inflorescence usually a dichasial cyme or thyrse, sometimes lax or subcapitate; bracts free, opposite, persistent, on the ultimate branchlets only 1 bract of each pair is flower-bearing; bracteoles present. Flowers hermaphrodite or unisexual, small, irregular. Calyx small, persistent; limb short during anthesis, inrolled, deeply divided into 10 or more segments, these unrolling in fruit and (in most species) developing into 5–15 plumose awns. Corolla funnel-shaped or campanulate, caducous after anthesis, small, gibbous, sometimes ± hairy in the throat; lobes (3, 4)5, oblong, patent, imbricate in bud. Stamens 3 (rarely 4, and occasionally 1–2 by abortion), inserted toward the top of the corolla-tube, usually exserted, alternating with the corolla-lobes; filaments thin; anthers small, 2-thecous. Style filiform, shortly 3-lobed or subentire, glabrous; stigma simple or 3-lobed. Fruit small, indehiscent, much compressed, with 3 dorsal, 1 ventral, and 2 marginal ribs.

Over 200 species, mostly found in the temperate and cold regions of the northern hemisphere, but occurring in all continents except Australia. Three species are known from East Africa.

Leaves crenate, sometimes pinnatifid towards the
    base;   inflorescence ± compact;   pappus
    obsolete .   .   .   .   1. *V. kilimandscharica*
Leaves generally pinnately divided, if undivided
    then entire or irregularly serrate; inflorescence
    ± lax; pappus present:
Plants clambering; stems branched at the nodes;
    stamens included .   .   .   .   .   2. *V. volkensii*
Plants not clambering; stems unbranched at the
    nodes; stamens exserted .   .   .   .   3. *V. capensis*

1. **V. kilimandscharica** *Engl.* in E.J. 19, Beibl. 47: 48 (1894); W.F.K.: 62 (1948); A.V.P.: 181–183, 330–331 (1957); Meyer in J.L.S. 55: 763, fig. 1 (1958). Types: Tanganyika, Kilimanjaro, above Kibosho Forest, *Volkens* 1533 (B, syn.†, BM, BR, G, K, PRE, isosyn.!) & Kilimanjaro, in the "Johannes-Schlucht", *Volkens* 1191 (B, syn.†)

Perennial rhizomatous subshrub 1–9 dm. tall, scrambling to reclinate or erect, ± tufted, rooting at the nodes from the creeping stems; rhizomes 1–10 mm. thick; stems rather profusely branched near ground level, leafy above, leafless below, woody, quadrangular in cross-section towards the top, terete below; internodes 0·3–2·0(–5) cm. long. Leaves mostly cauline, membranous, spathulate, 0·8–4·1 cm. long, 0·3–1·5 cm. wide in the flowering shoots, 2·5–5·7 cm. long, 0·7–1·4 cm. wide on the vegetative shoots, obtuse to acute, shallowly crenate to almost pinnatifid on the petiole below (lobes diminutive) or entire, pilose to glabrous. Inflorescence an aggregate dichasium, frequently globose to somewhat elongate, 0·8–4·8(–8) cm. long, 0·8–2·5 cm. wide; floral bracts cuneate-truncate, 2–8 mm. long, 0·8–3·0 mm. wide. Flowers nearly always perfect (plants occasionally gynodioecious), white, pink to purple or brownish-red. Calyx-limb short, cupuliform, ± dentate, not forming a feathery pappus in fruit. Corolla rotate, 5-lobed, barely gibbous, 1·5–2·5 mm. long, glabrous throughout; lobes regular or slightly irregular. Stamens 3, included; anthers 4-lobed, white. Style 3-lobed, included in perfect flowers, slightly exserted in pistillate flowers. Achenes broadly ovate to ovate-oblong, 1·7–2·6 mm. long, 1·0–1·7 mm. wide, glabrous (in specimens from Mt. Kenya and Kilimanjaro) or pubescent (in most specimens from Aberdare Mts. and Elgon), brown. Fig. 1.

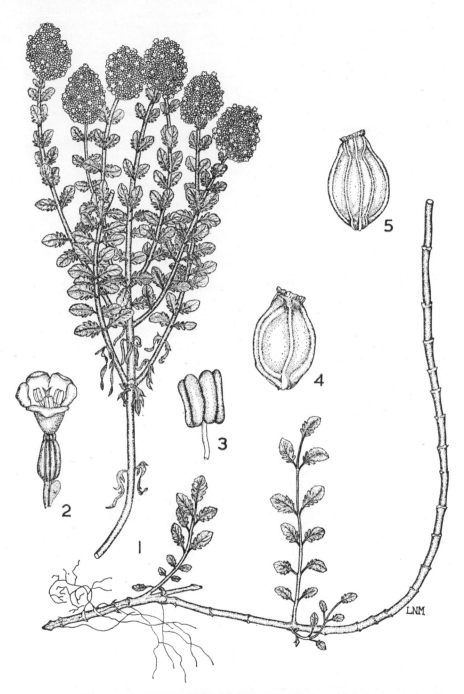

FIG. 1. *VALERIANA KILIMANDSCHARICA*—**1,** habit, × 1; **2,** flower, × 25; **3,** stamen prior to anthesis, × 50; **4, 5,** achene, in adaxial and abaxial view respectively. 1–3, from *Hedberg* 961; 4, 5 ,from *G. Taylor* 3503. Reproduced with permission from J.L.S. 55 (1958).

UGANDA. Elgon, Masaba, 6 Sept. 1933, *A. S. Thomas* 617!
KENYA. Elgon, NE. of crater, 16 May 1948, *Hedberg* 961!; Aberdare Mts., Kinangop, 13 July 1948, *Hedberg* 1549!; Mt. Kenya, Teleki valley, 26 July 1948, *Hedberg* 1699!
TANGANYIKA. W. Kilimanjaro, Shira Mts., Feb. 1928, *Haarer* 1140! & Kilimanjaro, near Peter's Hut, 22 Feb. 1943, *Greenway* 3749! & above Marangu in the *Philippia* region, 16 June 1948, *Hedberg* 1227!
DISTR. **U3**; **K3, 4**; **T2**; not known elsewhere
HAB. Upland moor, in short or tussock grassland and heath bogs, along streams and on wet rocks; 2800–4570 m.

SYN. *V. elgonensis* Mildbr. in N.B.G.B. 8: 234 (1922); Th. C.E. Fr. in N.B.G.B. 8: 413, t. 6/c (1923). Type: Kenya, Mt. Elgon, *Granvik* (S, holo.!)
V. keniensis Th. C.E. Fr. in N.B.G.B. 8: 412, t. 6/a (1923). Type: W. Mt. Kenya, alpine region, *Fries* 1379 (UPS, holo.!, BR, K, S, iso.!)
V. aberdarica Th. C.E. Fr. in N.B.G.B. 8: 413, t. 6/b (1923). Type: Kenya, Aberdare Mts., Sattima, *Fries* 2471 (UPS, holo.!, BR, K, S, iso.!)
V. kilimandscharica Engl. subsp. *elgonensis* (Mildbr.) Hedb., A.V.P.: 182 (1957)
V. kilimandscharica Engl. subsp. *aberdarica* (Th. C.E. Fr.) Hedb., A.V.P.: 183 (1957)

VARIATION. There are infraspecific variations in the achene indumentum, which tend to follow the trend that the populations on Mt. Kenya and Kilimanjaro have glabrous achenes while those on the Aberdare Mts. and Elgon have pubescent achenes. Sometimes this feature may vary within one collection, for example in *G. Taylor* 3503 (BM) we find both glabrous and pubescent achenes. There is also a tendency for the specimens from Elgon to have more pubescence on the leaves, and for those from the Aberdares to have narrower leaves with a larger number of crenules than the others (cp. Hedberg, A.V.P.: 331). The variations in all these features seem, however, too continuous to justify taxonomic segregation.

2. **V. volkensii** *Engl.*, P.O.A. C: 395 (1895); Z.A.E. 2: 342 (1911); Pl. Bequaert. 1: 288 (1922); Staner & Lebrun in Bull. Agric. Congo Belge 25: 427 (1934); A.V.P.: 181 (1957); Meyer in J.L.S. 55: 764, fig. 2 (1958). Type: Tanganyika, Kilimanjaro, above Useri, *Volkens* 1917 (B, holo.†, BM, BR, K, iso.!)

Perennial herb 5–10(–18) dm. tall, simple or branched, slender, rhizomatous and sometimes stoloniferous, often rooting at the nodes; stems herbaceous, usually clambering or erect, 1·5–5·0 mm. thick, generally leafless below and branched at the nodes above (first branches supported by leaves at the nodes); nodes solid. Leaves membranous, mostly cauline, imparipinnate, rhombic, largest at the base but decreasing in size and bract-like towards the inflorescence or undivided and ovate, glabrous to rather pilose; rhachis and the terminal leaflet of the undivided leaves 4·4–16·5 cm. long; apices and margins of the undivided leaves and the leaflets are characterized by glandular structures; terminal leaflet 1·8–6·5 cm. long, 0·6–3·4 cm. wide; lateral leaflets in 1–4 pairs, smaller than the terminal; undivided leaves ovate, 3–4 cm. long; petiole 0·5–4·8 cm. long. Inflorescence a dichasial thyrse 10–50 cm. long, 4–20 cm. wide; internodes glabrous or pilose; nodes tufted-pilose; floral bracts broadly lanceolate to ovate, 2·0–8·5 mm. long, 0·5–3·5 mm. wide. Flowers perfect (or ? occasionally imperfect), white to pale pink. Calyx-limb 8–13-fid, forming a feathery pappus in fruit. Corolla infundibuliform, gibbous, sometimes irregular, 1·5–3·0 mm. long; lobes one-third to one-half the tube-length; throat glabrous within. Stamens 3, included; anthers 4-lobed. Style barely 3-lobed and included, or distinctly 3-lobed and slightly exserted. Achenes elliptic to oval-patelliform, 2·5–4·5 mm. long, 1·5–3·0 mm. wide, glabrous throughout (in specimens from Kilimanjaro and Ruwenzori) or pubescent on the adaxial face (in most specimens from Elgon), fairly dark-brown. Fig. 2.

UGANDA. Elgon, Bamboo zone, Jan. 1918, *Dummer* 3577! & Elgon, 12 Nov. 1933, *Tothill* 2357!
KENYA. Elgon, E. side, 12 Dec. 1956, *Irwin* 313 & eastern slope above Tweedie's saw-mill, 3 Mar. 1948, *Hedberg* 257!

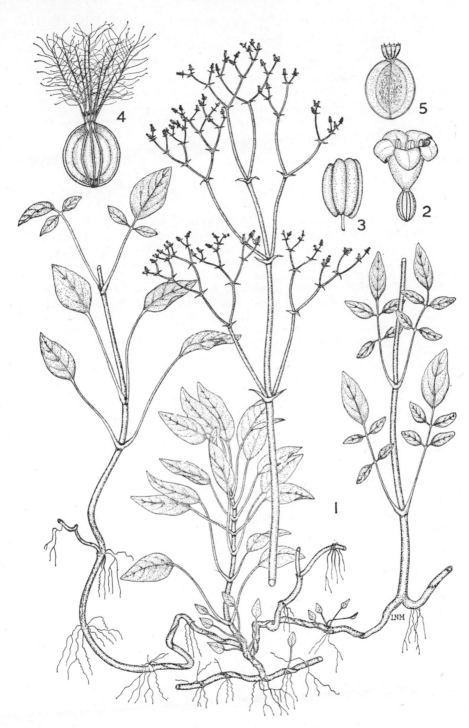

FIG. 2. *VALERIANA VOLKENSII*—1, habit, × 1; 2, flower, × 10; 3, stamen prior to anthesis, × 20; 4, achene, abaxial view, with feathery pappus, × 10; 5, achene, adaxial view, × 10. 1 from *Schlieben* 4823; 2, from *Alluaud* 155; 3, from *Synge* 1092; 4, 5, from *Synge* 1872. Reproduced with permission from J.L.S. 55 (1958).

TANGANYIKA. Kilimanjaro, Kimengalia stream above Rongai, 20 Feb. 1933, *C. G. Rogers* 438!

DISTR. **U3**; **K3**, 5; **T2**; Congo Republic

HAB. Moist ground along stream sides, in moist bamboo thickets, in upland rain-forest and in upland moor; 2300–3450 m.

VARIATION. Intraspecific variation occurs in the achene indumentum; in specimens from Kilimanjaro and Ruwenzori these are mostly glabrous, while the population on Elgon tends to have the achenes pubescent on the adaxial face. At least one collection (*Hedberg* 257 (S, UPS)) contains specimens of both types, however, so also in this species the pubescence of the achenes appears useless for taxonomic subdivision of the species.

3. **V. capensis** *Thunb.*, Prodr. Pl. Cap. 1: 7 (1784) & Fl. Cap: 145 (1811); Thonner, Blütenpfl. Afr., t. 146 (1908); Pillans in Flora of Cape Peninsula: 738 (1950); Meyer in J.L.S. 55: 766 (1958). Type: South Africa, Cape Province, near Lange Kloof, *Thunberg* (UPS, holo.!)

Perennial gynodioecious herb 1·2–8(–14) dm. tall in flower, 1·5–10 dm. tall in fruit; rhizomes 3–7 mm. thick; fibrous roots 2–4 mm. thick; sometimes stoloniferous; stems erect, herbaceous, leafy at the base, 1–4(–9) mm. thick, unbranched at the nodes (inflorescence sometimes branched but these branches supported by bracts); nodes solid and densely hairy, usually with shallow parallel ribs extending full length, terete to almost angled. Leaves membranous, predominantly imparipinnate but sometimes pinnatifid and occasionally undivided, with 1–9 regularly or irregularly arranged pairs of essentially sessile leaflets, mostly obovate to oblanceolate; basal ones loosely tufted; lowermost 7–44 cm. long, 1–10 cm. wide; cauline ones in 3–5 pairs, simulating the basal but smaller; uppermost bract-like; terminal leaflets ovate to lanceolate, 2·2–11·4 cm. long, 0·7–5·5 cm. wide, entire to dentate; lateral leaflets similar but smaller. Inflorescence a dichasial thyrse 1·5–30(–74) cm. long, 1·3–15(–21) cm. wide, subcapitate in the early stage; floral bracts linear-lanceolate, 0·6–2·0 cm. long, 0·7–2·5 mm. wide, acuminate; internodes glabrous or pilose along one side; nodes tufted-pilose. Flowers perfect or imperfect, white to pale pinkish-mauve. Calyx-limb 8–14-fid; segments forming a feathery pappus in fruit. Corolla infundibuliform, gibbous, sometimes irregular, 2·0–5·5 mm. long, densely to sparsely white hairy at the base of the throat within, glabrous without. Stamens 3, exserted (± twice the length of the corolla-lobes); anthers 4-lobed. Style unbranched at first, later opening into 3-lobed tip, exserted, generally the same length as the stamens. Achenes 2·5–5·5 mm. long, 1·2–2·5 mm. wide, glabrous, light-brown.

KENYA. Aberdare Mts., Cave Waterfall Gorge, 21 June 1962, *Coe* 757!; Mt. Kenya, NE. sector, Kathita R., 8 Aug. 1949, *Schelpe* 2622!

TANGANYIKA. Mbulu District: Mt. Hanang, Werther's Peak, 12 Feb. 1946, *Greenway* 7708!; Njombe District: Mdapo, Mar. 1954, *Semsei* 1703!; Songea District: Matengo Hills, Halau valley SE. of Miyau, 21 Jan. 1956, *Milne-Redhead & Taylor* 8228!

DISTR. **K3**, 4; **T2**, 7, 8; Mozambique, Malawi, Zambia and Rhodesia south to the Cape Province of South Africa

HAB. Upland grassland and upland moor, upland rain-forest at forest margins, by streamsides, in marshy ground or under overhanging dripping rock-faces; 1500–3400 m.

SYN. *V. capensis* Thunb. var. *lanceolata* N.E. Br. in K.B. 1895: 146 (1895). Types: South Africa, Natal, Tabamhlope Mt., *Evans* 368 (K, syn.!, MO, photo.) & Baziya Mts., *Baur* 546 (K, syn.!) & Malawi, Mt. Mlanje, *A. Whyte* (K, syn.!)

2. VALERIANELLA

Mill., Gard. Dict., Abridg. Ed. 4 [without pagination] (1754)

Annual (or biennial) herbs with dichasial branching, appearing dicho-
tomous, but the terminal bud abortive in the lower branches, glabrous or
pubescent. Radical leaves sometimes forming a rosette, entire, petiolate to
nearly sessile, undivided or dentate; cauline leaves often toothed, sometimes
entire, connate, rarely incised-pinnatifid. Inflorescence a terminal, usually
dichasial cyme subtended by lanceolate to oblong connate bracts. Flowers
sessile, solitary, hermaphrodite (or unisexual), in the axils of lanceolate to
oblong connate bracteoles. Calyx-limb often short or obsolete during flowering state, becoming hardened or variously enlarged in fruit, erect, spreading
or globose-inflated, regular or oblique, entire or 3–6-toothed, 3-horned or
divided into recurved or hooked rigid awns, never plumose-setose. Corolla
funnel-shaped, narrowly campanulate or tubular, attenuated at the base,
regular or sometimes gibbous, never spurred; limb 5-fid, spreading. Stamens
3, inserted towards the top of the corolla-tube, exserted; anthers 4-lobed.
Style shortly or minutely 3-fid at the apex; stigma simple or shortly 3-lobed.
Fruit an achene, oblanceolate to spathulate, glabrous to pubescent, 1-seeded,
3-locular, the 2 abaxial locules empty and ordinarily with a distinct groove
between them.

About 80 species, native of the temperate and subtropical countries of the northern
hemisphere, most numerous in southern Europe. One species is known from East
Africa.

The Corn Salad (*V. locusta* (L.) Betcke) is sometimes grown in agricultural experi-
mental stations of East Africa.

V. microcarpa *Lois.*, Not. Pl. Fr.: 151 (1810); De Nancy, Précis Trav.
Soc. Roy. Sc. Lett. Arts: 70, t. 1/6 (1833); Batt. & Tr., Fl. Alg.: 408,
t. 1/21 (1889); Rouy & Camus, Fl. Fr. 8: 97 (1901); Coste, Fl. Descr.
Illustr. Fr. 2: 268 (1903); Jahandiez & Maire, Cat. Pl. Maroc: 723 (1934);
Négre, Pet. Fl. Reg. Arid. Maroc Occid. 2: 240, t. 113/670 (1962). Type:
France, near Toulon, *G. Robert* (? P, holo.)

Annual slender nearly glabrous herb 5–25(–40) cm. tall; stems herbaceous,
erect or ascending to almost spreading, dichotomously and divaricately
branched, sometimes subsimple, weakly angled, slightly pubescent below,
1–3 mm. thick; tap-root 2·5–9 cm. or more long; secondary roots usually
slender. Leaves sessile, (0·5–)1·1–6·5 cm. long, 0·2–1·1 cm. wide, glabrous on
the surfaces, but sometimes ciliate at the base and on the margins; lower
ones entire, spathulate; upper ones dentate at the base, oblong, obtuse or
subacute; uppermost bract-like and linear. Inflorescence a compound
dichasium, glabrous; terminal internodes grooved and almost winged;
floral bracts lanceolate, 2·0–2·5 mm. long, 0·7–1·4 mm. wide. Flowers perfect,
sessile, pale pink to white. Calyx-limb obscure, oblique, scarcely veined,
entire (corolla, stamens, and style adhering to the achene-apex for some time).
Corolla funnel-shaped, barely saccate or attenuated at the base, glabrous
throughout; lobes 5, apparently regular but sometimes with 2 lobes smaller
than the other 3, 1·0–1·8 mm. long, generally pinkish above and whitish
below. Stamens 3, included or slightly exserted. Style at first club-shaped
and included, later very shortly 3-lobed at the tip and slightly exserted.
Achenes ovoid, 1–2 mm. long, 0·5–1·0 mm. wide, hispid to pubescent, or
glabrous (especially on the abaxial face), brown; fertile locule longer than
sterile ones, the latter not contiguous and appearing as linear ribs on the
abaxial face of the achene. Fig. 3, p. 8.

KENYA. Aberdare Mts., Oct. 1934, *G. Taylor* 1436!; Mt. Kenya, Sirimon Track, Dec.
1962, *Verdcourt* 3463! & Jan. 1963, *Verdcourt* 3529!

FIG. 3. *VALERIANELLA MICROCARPA*—1, habit, × 1; 2, part of inflorescence, × 7; 3, flower, × 25; 4, corolla, opened out to show stamens, × 25; 5, stamen, × 30; 6, ovary and style, × 25; 7, fruit, × 15; 8, same in transverse section, × 15; 9, seed, × 15. 1, from *Leaky & Evans* 144; 2–9, from *G. Taylor* 1436.

DISTR. **K3**, 4; Ethiopia, North Africa and Europe (Mediterranean region)

HAB. Upland moor (in Ethiopia also in upland rain-forest, bamboo thickets, grassland and persisting on cultivated ground); 2800–3200 m. (perhaps as low as 1000 m. in Ethiopia)

SYN. *Fedia microcarpa* (Lois.) Reichenb., Pl. Crit. 2: 6, t. 114 (1824)
     Valerianella abyssinica Fresen. in Mus. Senckenb. 2: 116 (1836); Verdc. in K.B. 21: 250 (1967). Type: Ethiopia, Simen Province, *Rueppell* (FR, holo.)
     [*V. dentata* sensu Hiern in F.T.A. 3: 248 (1877), *non* (L.) Poll.]
     V. dentata (L.) Poll. var. *microcarpa* (Lois.) Fiori in Fiori & Rochetti, Rel. Monogr. Agr.-col. 58: 40 (1940)

NOTE. With the limited amount of material available from the Flora area it is difficult to judge whether this plant is native or just naturalized. Records to date are from well-used mountain tracks and it may have been introduced, perhaps in mule fodder. On the other hand, being an annual, perhaps of short duration, producing small and rather inconspicuous flowers, it may simply have been overlooked and account for its poor representation in herbaria even if it is native to tropical East Africa (see also Verdcourt in K.B. 21 : 250 (1967)).

INDEX TO VALERIANACEAE